FLORA OF TROPICAL EAST AFRICA

ELATINACEAE

B. Verdcourt

Herbs or small shrubs, sometimes aquatic. Leaves opposite or verticillate, simple or submerged leaves sometimes much divided, entire or serrate; stipules present, paired. Flowers small, regular, hermaphrodite, axillary, solitary, cymose, fasciculate or in glomerules, occasionally cleistogamous. Sepals (2–)3–5(–6), free, imbricate, not nerved or 1-nerved, mostly with pellucid margins. Petals (2–)3–5(–6), hypogynous, imbricate, persistent. Stamens as many as and alternate to twice as many as the petals, free, hypogynous; anthers dithecous, opening by longitudinal slits. Ovary superior, (2–)3–5-locular, with axile placentation; ovules numerous; styles (2–)3–5, free. Fruit a septicidal capsule. Seeds straight or curved, without or with very little endosperm; embryo straight or curved, with short cotyledons.

A small family with only 2 genera and 35–40 species, occurring in both hemispheres in the tropics and in temperate regions. Both genera occur in Africa, but *Elatine* L. has not yet been found in the Flora area.

BERGIA

L., Mant. Pl. Alt.: 152, 241 (1771)

Prostrate, or less often erect, annual or perennial herbs or small shrubs, often glandular pubescent. Leaves opposite or pseudo-verticillate, linear to elliptic, entire or serrate, the stipules frequently denticulate or ciliate. Flowers axillary, sessile or pedicellate, solitary or fascicled or in verticillate glomerules. Sepals 3–5(–6), elliptic, oblong or lanceolate, often keeled with hyaline margins. Petals 3–5(–6), mostly oblong or elliptic and about the same size as the sepals. Stamens 3–12, mostly 5 or 10, those opposite to the petals sometimes shorter or absent, the rest with the filaments sometimes broadened at the base; anthers small. Ovary (3–)5-locular or the carpels almost free, each locule with numerous ovules; styles 3–5, free, usually shorter than the ovary. Fruit a 3–5-locular or 3–5-valved capsule or breaking up into distinct carpels. Seeds brown or almost black, oblong, subcylindric or 3-angled, sometimes slightly curved, faintly to strongly reticulate or tessellated; embryo straight or curved.

A small genus of about 20 species occurring in the tropics or subtropics of both hemispheres.

Herbaceous plants:
- Stems succulent; plant always glabrous; flowers in dense very compact verticillate glomerules; often perennial . . . . 1. *B. capensis*
- Stems slender, hard and wiry, not succulent; plant usually pubescent, rarely almost glabrous (not in Flora area); flowers in dense but laxer fascicles; annual . . . . . . 2. *B. ammannioïdes*

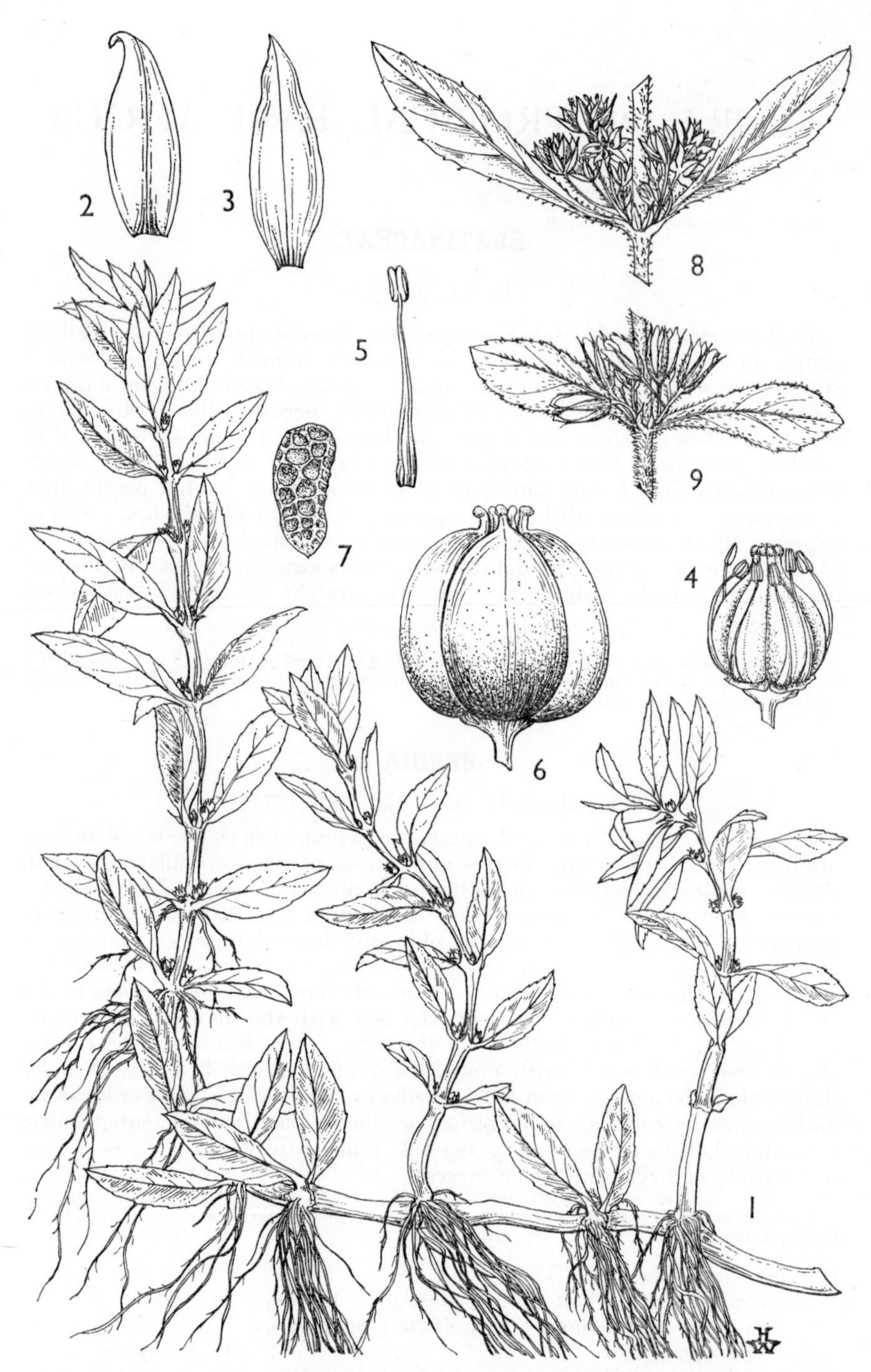

FIG. 1. *BERGIA CAPENSIS*—**1**, habit, × $\frac{2}{3}$; **2**, sepal, × 25; **3**, petal, × 25; **4**, flower with tepals removed to show stamens and gynoecium, × 20; **5**, stamen, × 25; **6**, fruit, × 20; **7**, seed, × 35. *B. AMMANNIOÏDES*—**8**, flowering node, × 2. *B. SUFFRUTICOSA*—**9**, flowering node, × 2. 1–7, from *Verdcourt* 2889; 8, from *Bogdan* 4443; 9, from *Bally* 5887.

Woody spreading shrublet with densely scabrid-pubescent stems and leaves . . . . . 3. *B. suffruticosa*

1. **B. capensis** *L.*, Mant. Pl. Alt.: 241 (1771); Milne-Redh. in K.B. 3: 450 (1949); F.P.S. 1: 83 (1950); Backer in Fl. Males. 4: 203, fig. 1 (1951); Keay, F.W.T.A., ed. 2, 1: 128 (1954). Type: type-locality erroneously given as Cape of Good Hope, but presumably the specimen came from Asia, specimen 597.1 (LINN, lecto.!)

Glabrous herb with thick creeping succulent stem, which roots at the nodes, up to 50 cm. long; branches erect or ascending, ± 7–30 cm. tall, often reddish. Leaves opposite, subsessile or shortly petiolate; blade oblong-lanceolate or lanceolate, 1·3–5 cm. long, (0·4–)0·75–2·5 cm. wide, acute or obtuse at both ends, finely serrulate, the teeth reddish; petiole 1–5 mm. long; stipules ovate-triangular, 2–4 mm. long, acute or acuminate, membranous, dentate. Flowers 5-merous, many in dense clusters; pedicels 0·5–3 mm. long. Sepals lanceolate to broadly elliptic, 1–2 mm. long, acuminate or mucronulate, green tipped with red. Petals at first erect, later spreading or recurved, oblong, lanceolate or subspathulate, slightly longer than the sepals, 1–2·2 mm. long, 0·5–0·6 mm. wide, white or pink, transparent. Stamens 10; filaments thin, filiform, slightly widened at the base. Ovary subglobose; styles recurved, 0·25–0·3 mm. long. Capsule crimson, subglobose, 2·5 mm. in diameter, with 5 longitudinal furrows. Seeds oblong-ellipsoid, often curved, brown, shining, ± 0·5 mm. long, strongly reticulate. Fig. 1/1–7.

TANGANYIKA. Mwanza District: Mwanza, *R. L. Davis* 234! & 1·6 km. from Busisi–Kikongo Ferry on Geita road, near Lake Victoria, 17 July 1960, *Verdcourt* 2889!; Ulanga District: near junction of Kilomberu, Luwegu and Rufiji rivers, 14 June 1932, *Schlieben* 2390!

DISTR. T1, 6; Gambia, Ghana, Egypt, Sudan Republic*, Rhodesia, ? South West Africa; also Iraq, Caucasus (Talysh), India, Ceylon, Vietnam, Java (see note)

HAB. Swampy areas, rice-fields, shallow pools, river-banks and damp grassland; in Flora area in pools near lake-shore with *Eclipta*, *Ammannia*, *Commelina*, *Nesaea*, etc.; 1140 m.

SYN. *B. verticillata* Willd., Sp. Pl. 2: 770 (1799); Oliv., F.T.A. 1: 152 (1868), *nom. illegit.*

NOTE. Additional synonymy not concerning the Flora area is given in the references cited. It appears to me that *B. sessiliflora* Griseb., described from Cuba, is the same species and a single specimen from Ecuador also appears to belong here. If so its presence in the New World is due to introduction.

2. **B. ammannioïdes** *Roth*, Nov. Pl. Sp.: 219 (1821); Oliv., F.T.A. 1: 152 (1868), as "*ammanoides*"; Engl., V.E. 3(2): 523, t. 237/K–R (1921); Niedenzu in E. & P. Pf. 21: 274, fig. 119/K–R (1925); F.P.S. 1: 85 (1950); Backer in Fl. Males. 4: 205 (1951); Keay, F.W.T.A., ed. 2, 1: 128 (1954); Wild in F.Z.1: 373, t. 72/A (1961), pro parte; Boutique in F.C.B., Elatin.: 2 (1967). Type: India, *Heyne* (location not known, holo.)

Annual, erect or somewhat decumbent unbranched or branched herb, or prostrate with erect shoots, 8–30(–50) cm. tall or long, mostly pinkish, hard, usually pilose with capitate or glandular or sometimes woolly hairs but sometimes almost glabrous in Asian specimens. Leaves opposite, subsessile or shortly petiolate (mostly very short in African material); blade elliptic-oblong, oblanceolate, oblong or obovate-oblong, 0·4–5 cm. long, 0·2–2 cm. wide, acute or blunt at the apex, cuneate, serrulate to almost entire, sometimes remotely glandular ciliate, glabrous or sparsely hairy above, glandular

* Specimen cited in F.T.A. 1: 152 (1868) not seen, apparently missing.

pubescent or with filamentous hairs beneath, rarely almost glabrous; petioles 2–8 mm. long; stipules narrowly triangular, lanceolate or subulate, 2–3(–5) mm. long, acute, membranous, glandular hairy with ciliolate or serrulate margins. Flowers 3–5-merous, many in dense clusters or few to many in dense to fairly loose fascicles; pedicels slender, 1–5 mm. long, glandular pubescent, mostly very short in African material. Sepals lanceolate to ovate-oblong, 1–3 mm. long, acute or distinctly acuminate, the keel and one or both margins often ciliolate, or entire, the surface glandular pubescent. Petals lanceolate or oblanceolate, slightly shorter than or equalling the sepals, 1–2·5 mm. long, 0·4–1(–1·25) mm. wide, obtuse or subacute, white or pink, transparent. Stamens (3–)5–10(–12), when more than 5 then alternate ones slightly longer and broader. Ovary ovoid or globose; styles recurved, 0·2–0·3 mm. long. Capsule reddish, ovoid, 1–2 mm. long. Seeds subcylindric with rounded ends, dark brown, shining, 0·3–0·4 mm. long, minutely tesselated or faintly reticulate. Fig. 1/8, p. 2.

KENYA. Embu District: 29 km. SSW. of Embu, Mwea-Tebere Irrigation Station, 20 Feb. 1957, *Bogdan* 4443! & 12 Aug. 1958, *Bogdan* 4612!

TANGANYIKA. Dodoma District: Great North Road, 25·6 km. N. of Dodoma, 21 Apr. 1962, *Polhill & Paulo* 2115!; Iringa District: where Great North Road crosses the Great Ruaha R., Mtera, 19 Apr. 1962, *Polhill & Paulo* 2076!

DISTR. **K**4; **T**5, 7; W. Africa, Sudan Republic, Mozambique, Zambia, Rhodesia, South West Africa; also Arabia, Iraq, Iran, Afghanistan, W. Pakistan, Uzbek S.S.R., India, Ceylon, Thailand, Vietnam, southern China, Malesia to Philippines, Australia (see note)

HAB. Clay-flats with scattered *Acacia*, *Cordia*, etc., also as a weed in rice-plots in shallow water in irrigation trials recently prepared from former grassland; in general throughout its range in moist places and by dried up pools; 780–1110 m.

NOTE. Synonymy not concerning the Flora area is given in the references cited. I do not agree with Backer that *B. serrata* Blanco (*B. glandulosa* Turcz.) is completely synonymous with *B. ammannioïdes*; although similar the flowers are much larger and the pedicels up to 2·2 cm. long. I also do not agree with Wild that the glabrous forms he refers to *B. ammannioïdes* truly belong to that species; they are I think extreme variants of *B. capensis*. *B. ammannioïdes* is very variable nevertheless, and the variation is not well correlated with geography. Long range dispersal by birds may afford an explanation. Most of the material from Australia referred to this species appears racially distinct.

3. **B. suffruticosa** (*Del.*) *Fenzl* in Denkschr. Bot. Gesell. 3: 183 (1841); Oliv., F.T.A. 1: 153 (1868); Engl., V.E. 3(2): 525, fig. 237/A–J (1921); Niedenzu in E. & P. Pf. 21: 274, t. 119/A–J (1925); F.P.S. 1: 83, fig. 56 (1950); Keay, F.W.T.A., ed. 2, 1: 127, fig. 45 (1954). Types: Egypt, Nile, Sa'yd, Gebel Selseleh, Qoubbanyeh, *Delile* (MPU, syn.) & Sudan Republic, Nile between Blocho and Dongola, *Lippy* (ubi?, syn.)

Aromatic woody-based spreading shrublet with numerous opposite branchlets, the erect shoots 5–20(–60) cm. tall; plant at first densely shortly hispid, the older stems at length almost glabrous with the bark papery and peeling in rust-coloured flakes. Leaves opposite or pseudo-verticillate, usually rather thick, sessile or shortly petiolate; blade elliptic or oblanceolate, 0·2–3 cm. long, 0·1–1 cm. wide, obtuse to acute at the apex, cuneate, crenate or serrate, usually markedly so, margins sometimes revolute, mostly densely scabrid-pubescent, sometimes glandular; petiole 0–3 mm. long; stipules linear or lanceolate, 1·5–2·5 mm. long, mostly not membranous, pubescent. Flowers 5-merous, solitary or in 2–8-flowered fascicles; pedicels 1–5 mm. long, pubescent. Sepals ovate, 2·5–5 mm. long, acute to distinctly acuminate, keeled, pubescent, usually with membranous margins. Petals obovate-oblong, ± equalling the sepals, 3–4 mm. long, 1·2–2 mm. wide, obtuse, white to pinkish-mauve, thin. Stamens 10, nearly equal or alternately shorter, those opposite the sepals often dilated below. Ovary ovoid, 5-sulcate; style

1–1·2 mm. long. Capsule pinkish, ovoid, 2 mm. long. Seeds oblong-ellipsoid, dark brown to black, shining, 0·5–0·6 mm. long, faintly reticulate. Fig. 1/9, p. 2.

KENYA. Northern Frontier Province: Wajir, 18 Jan. 1955, *Hemming* 487! & 4·8 km. from Garba Tula on Isiolo road, 18 July 1952, *Bally* 8226! & Tanaland, Derisa, 22 Aug. 1946, *J. Adamson* 401!; Northern Frontier Province/Tana River District boundary: Bura, 17 Mar. 1963, *Thairu* 76!

DISTR. **K**1, 7 (north); Senegal and Mauritania to Egypt, Sudan Republic, Arabia, W. Pakistan and western India

HAB. Hard seasonally wet pans with *Sphaeranthus*, *Glinus*, *Lawsonia*, *Aerva*, etc.; 100–750 m.

SYN. *Lancretia suffruticosa* Del., Fl. d'Egypte: 69 [213], t. 25/1 (1813)

INDEX TO ELATINACEAE